FROM THE PROFOUNDNESS OF DREAMS

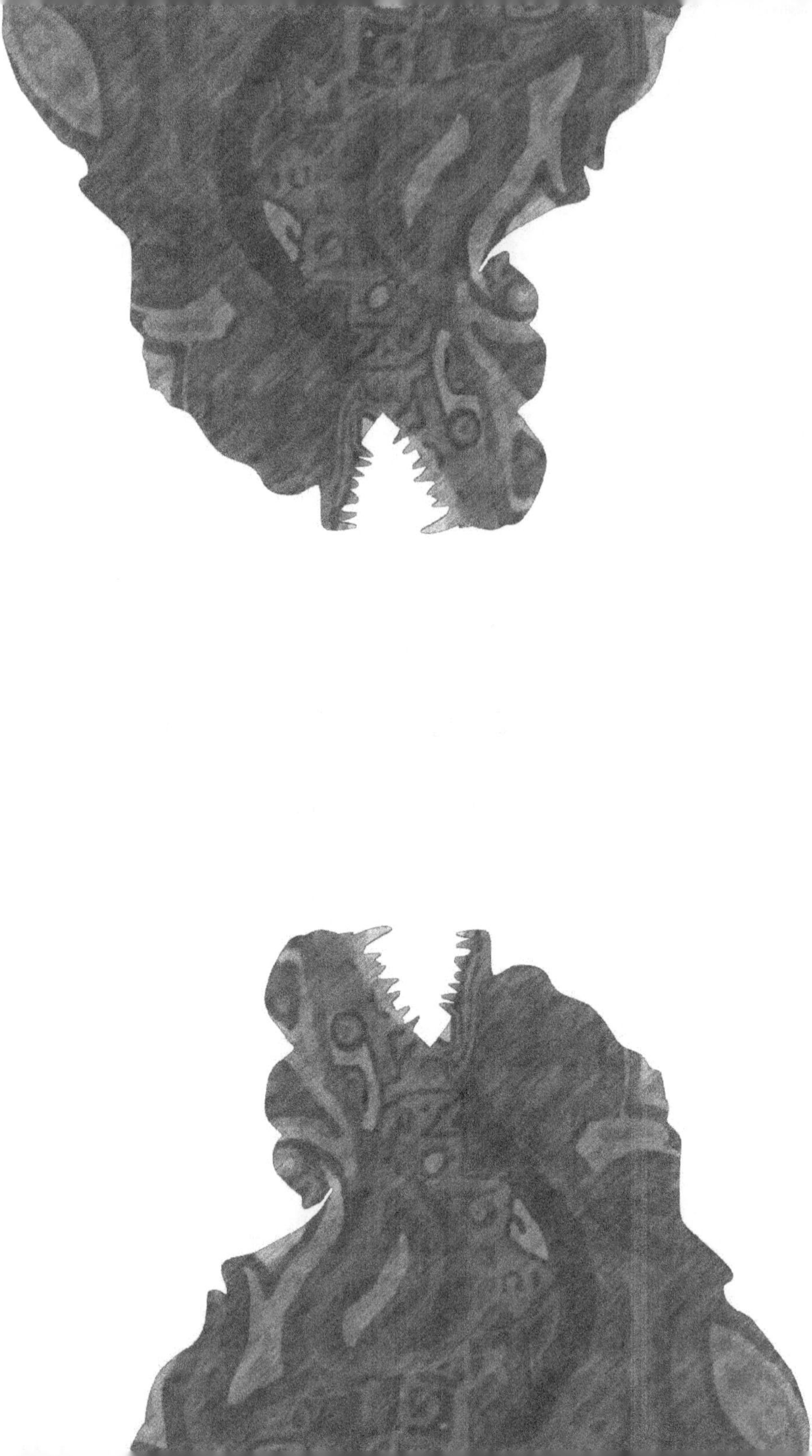

From the Profoundness of Dreams

An Analysis of Sleep, Dreams, and their Relation to Reality

Written by Daniel Strauss

Artwork by Tobias Wegener

With a foreword by Wolfgang Vincke

Title: From the Profoundness of Dreams
Subtitle: An Analysis of Sleep, Dreams, and their Relation to Reality

Author: Daniel Strauss
Cover artwork: Tobias Wegener
Foreword by: Wolfgang Vincke
Published by: Daniel Strauss
Distributed by: Lulu Enterprises

This book can be found on the Lulu Store and on Amazon.

ISBN 978-1-291-13066-9

Acknowledgments

I sincerely thank Tobias Wegener, who designed the unique cover for this edition and transformed it into something dreamy and beautiful.

I also deeply thank Wolfgang Vincke, author of "blicken – Vier Erzählungen", who wrote an inspiring foreword for this book, creating a pleasant and interesting atmosphere for the rest to come. He supported me until the end, reading both of the essays beforehand as well as affirming my idea to captivate them in a single volume. I am very grateful for his efforts.

"Your visions will become clear only when you can look into your own heart. Who looks outside, dreams; who looks inside, awakes."

- C. G. Jung

Foreword

"In college we barely talk about the subject and if we do, we never look at dreams with the profundity that I would expect."

One day this year, Daniel handed an essay to me in which he scrutinizes his personal fascination with dreams, titled "Subjective Realities from Within. A Journey to the Edge of Consciousness." What makes a student complain about the lack of interest in this topic in class? And what makes a teacher accept a couple of weeks later the student's ambitious request to write his scientific essay about dreams, in particular about lucid dreams? Actually I don't know. Yet Daniel and I feel united by the common interest in the adventure of creativity and imagination and thus I was looking forward to an essay that takes you to get to your inner world. Daniel successfully has combined his talents to create a short and surprising work that explores the nature of creativity and takes you beyond the popular physics fact that sometimes you cannot tell whether you are awake or asleep. After reading, I seriously think about a method to permanently integrate this peculiar issue into my English lessons, in a way the philosopher Sir Thomas Browne once had described his experience with it: "… yet in one dream I can compose a whole Comedy, behold the action, apprehend the jests and laugh myself awake at the conceits thereof".

Wolfgang Vincke
Author of "blicken – Vier Erzählungen"
June 2012

Preface

The initial idea for this book was to captivate my knowledge about dreams up to this point in a single volume. This book includes not only a non-fictitious essay but also one that is indeed fiction. "From the Profoundness of Dreams" is the first essay, being of a non-fictitious nature and summarizing my factual knowledge on the subject. This book bears the title of that essay as it is the main content within this volume. The second essay entitled "Subjective Realities from Within" is a short story in which my fantasies are let lose, demonstrating how far the unconscious mind can take you from the world one claims to know.

The book is divided into two parts, each part entitled after the respective essay. This book holds everything I have written so far on the subject of dreams and I truly desire for it to be appreciated as such.

Daniel Strauss
June 2012

From the Profoundness of Dreams

Daniel Strauss

The Symbol seen above is Japanese and means "Dreams"

Table of Contents

Part I

From The Profoundness of Dreams

An analysis of sleep, dreams, and their relation to reality

Introduction

Dreams are mostly still a mystery to us all. Nobody knows where they come from nor does anyone know where their meaning lies. We live in a world of technology, in a world where time seems to move faster than we can handle. Most beings take no time to care or solve mysteries as such.

Yet we must ask ourselves what is important, we must set priorities in order to manage our time efficiently. Most do not make thinking about dreams a priority, yet I, among many other famous beings such as Freud and Jung, believe that by doing so we can learn to understand ourselves and ultimately lead a better life. Dreams are a product of our very own nature, unstained by pollution and technology. Humanity should worship as they should explore the little nature that is left within us.

My goal in this essay is to introduce the concept of dream analysis, using the methods and theories of psychologist Carl Gustav Jung, as well as the phenomena referred to as lucid dreaming. I will also discuss the main problems in the field of dreams and how those are to be treated. Ultimately, I wish to underline how spending time with your dreams can enrich your life, drawing lines between the waking world and the dream world, as well as between famous personalities in the field of dreams.

1. Dreams

Dreams are an alternate model of our world which we experience in many different ways every night. This alternate model is produced during sleep by an apparatus of our mind referred to as the unconscious, based on past experiences, memories and external stimuli. The Oxford English dictionary defines a dream as a "train of thoughts, images, or fancies passing though the mind during sleep"[1], yet as one can assume from personal experiences, this definition while being strictly scientific, fails to captivate the emotional realism we experience in our dreams. Carl Gustav Jung defined a dream as the direct expression of unconscious psychic activity[2], while Alphonse Maeder[3] stated that a dream is a spontaneous self-portrayal, displaying the current emotional state of the unconscious[4]. All definitions relate, as they describe unconscious psychic activity showing us things of relevance through an environment very different from our own.

1.1 The Manifestation of a Dream

Knowing what a dream is, we must ask ourselves which factors determine what we dream and how such a dream makes its way into our memory. A dream, without doubt, is a psychic product, created by our unconscious. The diagram M1, seen on the next page, demonstrates how this product makes its way into our conscious memory and which factors determine its existence. The rawest manifestation of a dream is called the **latent dream content** and is influenced by **five primary factors**. The most natural influences are

---

[1] "Lucid Dreaming" by Stephen LaBerge, Chapter 2, Publisher: Sounds True, 2009 Edition, Chapter 2, p.10
[2] "The Practical use of Dream Analysis" by C. G. Jung, from "Dreams", Bollingen Series, 2010 Edition, p. 88
[3] Alphonse Maeder was a Swiss physician as well as assistant of Jung.
[4] "General Aspects of Dream Psychology" by C. G. Jung, from "Dreams", Bollingen Series, 2010 Edition, p. 49

M1 Dream Manifestation[5]

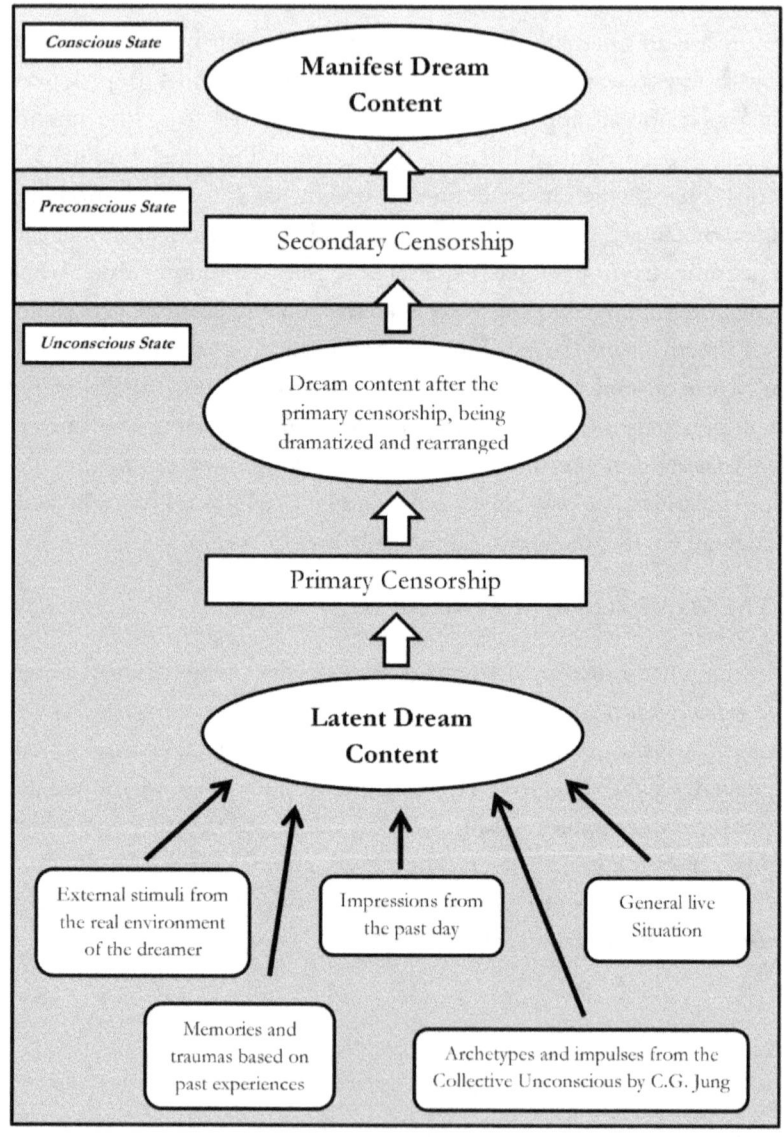

---

[5] Adapted from "Schöpferisch Träumen" by Paul Tholey and Kaleb Utecht, publisher Klotz, fifth edition, 2008, page 11.

stimuli from the outside world at the time of sleep, referring to smells, sounds as well as other sensations perceived by the dreamer. The sound of a train or of a bird may be handled by our unconscious mind to display images that somehow correlate with the external input. The usage of these external sensations to modify the dream world is naturally kept at a minimum, as the mind is more focused on inner sensations such as emotions rather than on our external senses during a dream[6].

The memories of the past day also play a crucial role as they are very present in the dreamer's mind and are still to be processed by our psyche. A person who just got a promotion for example, is more likely to dream something about that particular happening, as it is very important, yet minor happenings such as a beautiful tree that was observed by a person on the way home may also find its way into the dream even if the person at the time of occurrence did not even consciously realize that the tree was there. The unconscious mind comprehends things that our conscious mind sometimes does not even fully take into account.

The general life situation of the dreamer is also very relevant because it determines how the dreamer's emotional situation is at the current time, influencing the process of dream manifestation.

The factor that is definitely more complex to deal with captivates **past traumatic experiences**, such as fear from heights, fear from a certain animal such as spiders, as well as sexual traumas of similar relevance. While being awake our conscious mind usually suppresses these traumatic memories, leaving them to be unfolded in our unconscious mind.

The last of the five primary factors is one that may not be recognized by some psychologists, yet it is one that explains many collective fears

---

[6] "Exploring the World of Lucid Dreaming" by Stephen LaBerge and Howard Rheingold, Random House publisher, 1991 Edition, Chapter 5, p. 127

and desires across humanity: **Jung's archetypes and the collective unconscious**, to be dealt with in depth later in this essay.

At this point the **latent dream content** is present in the unconscious, influenced by the mentioned factors. The process this latent content undergoes to make its way into our memory, per say into our conscious mind, is full of censorships, as well as alterations. The product makes its way to the edge of our unconscious undergoing a primary censorship, reordering the dream content as well as dramatizing it to be captured by our conscious mind, in an altogether more accessible way. Before reaching our conscious mind as the well-known foggy dream memory, the content undergoes a secondary censorship, in which the dreamer already makes assumptions and connections in a preconscious state. Ultimately we wake up, regaining full consciousness and having the dream memory, referred to as the **manifest dream content**, present in our conscious mind[7].

To be emphasized is the fact that this model of the manifestation of a dream cannot be applied to the real process in this exact manner. The model is merely a generalization of the process for better visualization and therefore understanding of the procedure. The crucial point is that a dream as we remember it, the **manifest dream content**, has been altered greatly from its raw substance, the **latent dream content**, making analysis much more complex and difficult.

1.2 The Architecture of a Dream

"Dreams are models of the world" – Stephen LaBerge

At this point it is necessary to gain an even deeper insight at how dreams are constructed within our unconscious mind. There are many theories that deal with this essential question, yet the one that is mostly recognized by modern science comes from the lucid dreaming pioneer

---

[7] "Wie ein Traum entsteht" by Paul Tholey and Kaleb Utecht, from "Schöpferisch Träumen", publisher: Klotz, fifth edition 2008, p. 10-17

Stephen LaBerge. Previously we established that the latent dream content is influenced by five primary factors, meaning that they decide what we dream, yet we do not know what is precisely happening within our unconscious mind at the time of dream creation.

LaBerge answers this question with **schemas**, being mental models of entities that allow us to recognize objects, emotions, values, as they are. When we perceive something our brain is looking for a way to recognize the perceived, for example a table or an ocean. Our brain is able to do so by relying on schemas. There are schemas for everything, from simple objects like a chair or a table to complex emotions and values like love and justice. While being awake our brain is constantly recognizing objects and situations based on what we perceive with our senses, in other words, schemas are constantly being activated, pulled into our conscious mind. The result of schema activation is a conscious experience. We create new schemas every day, by combining existing ones or discovering something entirely new[8].

I wish to demonstrate how schemas operate with the following example: think about the moon. Right now the schema for the moon is in our conscious mind. The fact that the schema of the moon has been activated only means that several other relatable schemas are now laying in the preconsciousness of our mind, such as stars and planets. Now that they have been mentioned, they are in our conscious mind as well, with all their characteristics and attributes. Just a second ago schemas like music or food were probably not activated at all, lying in the unconscious, now naturally they are activated since they have been mentioned. In the moment where the word moon, star or planet was being read, our brain searched in an instant for a way to categorize this input from our perception, recognizing several schemas and pulling them above the unconscious along many others that relate to the mentioned ones. Our brain is forced to process a huge amount of infor-

---

[8] "Exploring the World of Lucid Dreaming" by Stephen LaBerge and Howard Rheingold, Random House publisher, 1991 Edition, Chapter 5, p. 122

mation every second of our lives, meaning that there has to be some kind of system, some patter of organization that the brain uses to remember and categorize what things are. LaBerge's theory of schemas is therefore a comprehensible approach to understand how perception, categorization and ultimately identification of the outside world works.

Schemas are just as relevant to us while being awake as they are while being asleep. In the awakened state the primary input to determine which schemas are activated comes from our senses, what we feel, hear, smell, taste and see. LaBerge states that when we are dreaming, the brain barely receives any sensory input from the outside world, having to mainly rely on what is within our mind to activate schemas and build the dream world[9]. The previously mentioned five primary factors determine the latent dream content, in other words, they determine which schemas are activated to create the dream world. This leads us to LaBerge's Final thesis, dreams are models of the world, build by the schemas we have within our mind.

1.3 The Architecture of Sleep

Knowing how sleep works and at what part of the night dreaming occurs is the final amplification to be made before proceeding to advanced topics such as analysis and lucid dreaming. Sleep is primarily divided into two main types: Rapid-eye movement, referred to as **REM** and non-rapid eye movement, referred to as **NREM**.[10]

NREM consists of **four stages** referred to as **N1, N2, N3** and **N4**. **N1** is the first stage of sleep, characterized by drowsiness and occasional twitches, lasting five to ten minutes. In the **N2** stage, muscle activity, temperature and breathing decreases. Conscious awareness of the outside world begins to vanish, leaving the body in a state of relaxa-

---

[9] "Exploring the World of Lucid Dreaming" by Stephen LaBerge and Howard Rheingold, Random House publisher, 1991 Edition, Chapter 5, p. 128
[10] "Sleep", Wikipedia, URL: http://en.wikipedia.org/wiki/Sleep (02.04.2012)

tion, ready to enter the **N3** stage. **N2** only lasts a couple of minutes[11]. The **N3** stage is often referred to as delta sleep or deep sleep, since brainwave activity is reduced to a minimum and muscles are completely relaxed. Going into the depths of brainwave activity is outside of the scope of this essay, yet M2 seen below, demonstrates a chart where one can see how brainwave activity is reduced in each deeper sleep stage.

M2 Brainwave Activity[12]

Stage	Frequency (Hz)	Waveform type
awake	15-50	-
pre-sleep	8-12	alpha rhythm
N1	4-8	theta
N2	4-15	spindle waves
N3	2-4	spindle waves and slow waves
N4	0.5-2	slow waves and delta waves
REM	15-30	-

N4 is an expansion of **N3** in which the body falls into an even deeper sleep. In many illustrations **N4** and **N3** have been combined into **N3** because there is little difference between the two sleep stages. **N3** respectively **N4** lasts approximately 30 minutes. In the **REM** stage, dreaming occurs, characterized by rapid eye movements, occasional twitches, rapid breathing, intensified heart rate and Sleep paralysis, also referred to as **SP**. **SP** is a state of paralysis that the body induces slightly before the **REM** stage to insure that the dreamer does not act out his dreams. Surprisingly in the **REM** stage, as observed in M2, the brain activity is at its highest, just like while being awake.

---

[11] "Sleep Stages" by Thiruvelan, URL: http://healthy-ojas.com/sleep/sleep-stages.html (02.04.2012)
[12] Adapted from "Sleep Stages" by Thiruvelan, URL: http://healthy-ojas.com/sleep/sleep-stages.html (02.04.2012)

M3 Hypnogram[13]

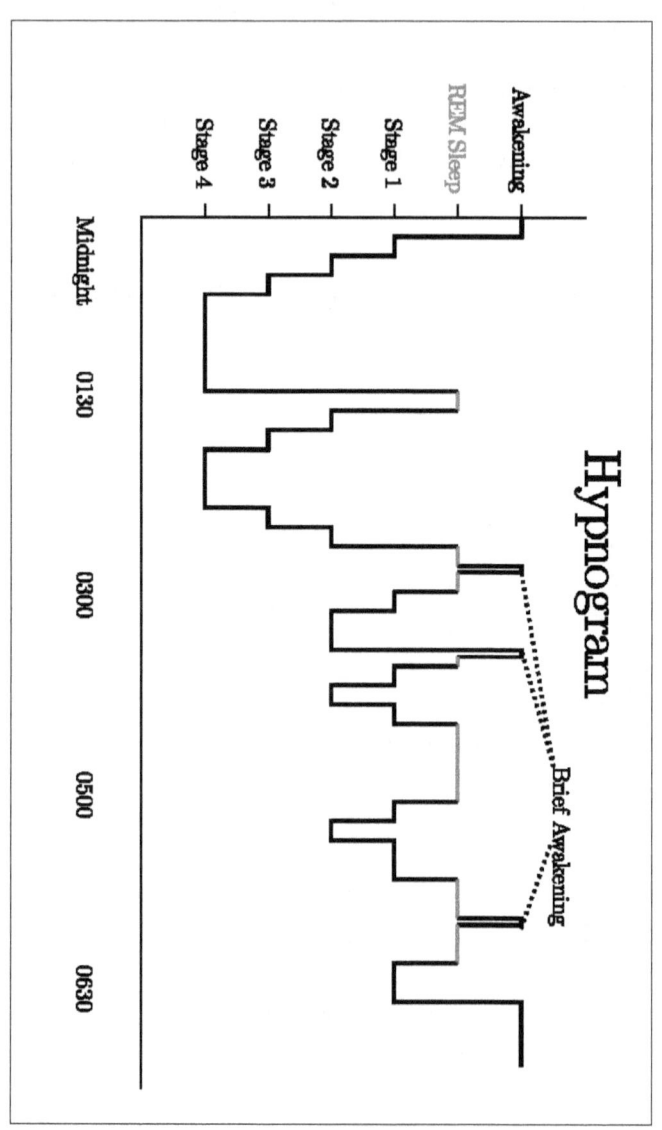

---

[13] "Sleep_Hypnogram.svg", Wikipedia, URL:
http://en.wikipedia.org/wiki/File:Sleep_Hypnogram.svg (02.04.2012)

The **N2** stage, while being very short but occurring often during the night, and the **REM** stage, are the stages where we spend most time in during sleep[14]. These sleep stages do not happen in chronological order as one might think.

M3, seen on the previous page, demonstrates a hypnogram[15] presenting the usual sleep pattern of a human being.

On the **x axis** the time is labeled, from midnight to 6:30 AM. On the **y axis** the Sleep stages are labeled, starting with the stage of being awake, all the way down to **N4**, the deepest sleep stage. The hypnogram clarifies that we sleep in cycles, each lasting around **90 minutes**, usually culminating in the **REM** stage. As the night proceeds each **REM** phase becomes longer, with the first one lasting around ten minutes, up to the last one being able to last around fifty minutes. What also becomes immanent is that **REM** sleep is the lightest sleep we could possibly find ourselves in, resulting in small awakenings every now and then, yet they are so minor that the average human being does not even take notice of them. As the night progresses we spend less time in **N3** and **N4** stages, and more time in the **REM** stage. The order in which our body goes through the sleep stages is reversed when moving into the next **REM** phase of the night, going from **N4** not immediately to **REM** but rather to **N3**, following **N2** and sometimes even **N1** again, before finally proceeding into the next **REM** phase. The normal succession of the sleep stages is visualized in M4 seen below.

M4 Normal Succession of the Sleep Stages[16]

---

[14] "How Dreams Work" by Lee Ann Obringer, URL:
http://science.howstuffworks.com/environmental/life/human-biology/dream2.htm (02.04.2012)
[15] A hypnogram is a sleep profile of a human being.
[16] Adapted from "Sleep Stages Overview, Sleep Cycle" by Stanley J. Swierzewski, URL:
http://www.hc.com/sleep-stages/overview-sleep-cycle.shtml (09.04.2012)

Knowing how sleep works, what happens in the **REM** stage and how dreams manifest, is the basis to understand how we can make use of this knowledge, in other words, how we can analyze the succession of images that we are confronted with every night to achieve better self-understanding, leading us to C. G. Jung's psychoanalysis.

2. Carl Gustav Jung and his Psychoanalysis

Carl Gustav Jung was born 1875 in Switzerland and lived a very troublesome childhood, preferring to be alone with his thoughts and experiencing his first encounter with neurosis. He would faint every time he had to do anything related to academics. Later he managed to overcome the problem and develop a renewed interest in academics. Jung's primary interests lied in medicine and spiritual phenomena, which led him into the field of psychiatry. After graduating from university with a medical degree in 1902, he got married to Emma Rauschenbach until she passed away in 1955. In 1906 he shared some of his work with psychologist's founding father Sigmund Freud. His positive impression was the beginning of a friendship between them that lasted for several years. When they finally met in person they had a conversation that lasted for over twelve hours without pause. The friendship with Freud had a major impact on Jung's theories and awoke an intense interest for the unconscious mind within him. Eventually, Jung's theories began to separate from Freud's as he developed new ideas that rejected Freud's emphasis on sex as the essence behind all dreams. Freud was disappointed as he saw in Jung his possible protégé. Parting from Freud was not easy, many colleagues and friends turned against him as Freud closed ranks among his followers. In the following years Jung became fascinated with dreams and symbols as he founded his own theory called analytical psychology. In the six year period that followed, Jung suffered from psychosis as he embarked on an unforgettable journey to explore his own unconscious, recording his experiences in a previously unpublished book known as The Red Book. The Red Book was published several years after his death in 1961, setting a milestone in the field of psychiatry[17].

---

[17] "Carl Jung Biography" by Kendra Cherry, URL:
http://psychology.about.com/od/profilesofmajorthinkers/p/jungprofile.htm
(02.05.2012)

A dream can have many meanings and interpretations, it is important to emphasize that analyzing a dream is by no means a linear process. The analyst must take into account many possibilities and interpretations, and ultimately decide which one delivers the most sense making result. Over the years Jung analyzed hundreds of dreams, recognizing certain patterns and developing tools and approaches that ultimately formed his theories about dream analysis. In the following I will demonstrate his theories and make use of them to analyze a dream based on a real-life example.

When dealing with a psychic product such as dreams, one must make a classical distinction, in other words, every psychic product has a two-fold point of view; **causality** and **finality**. By **finality** the question as to what purpose does it happen is meant, while **causality** defines the question as to why it happens. Both finality and causality should be taken into full account to arrive at a productive result[18]. When confronted with a dream of a patient one must, before jumping to any conclusions, gather information to **establish the context**[19]. This is a crucial step, as it defines the further proceedings of the analysis. The analyst must be aware of the conscious situation of the dreamer; he must know what relation exists between the dream images and possible experiences in the past[20]. This procedure is not simple; one might easily wander off and gather information that is unnecessary to the topic at hand. To avoid this requires experience, Jung quotes the famous philosopher Immanuel Kant to illustrate how this can be achieved, explaining that to "comprehend" a thing means to "cognize it to an extent necessary for our purpose"[21]. Jung developed a typical question to ensure the correct accumulation of information. He would ask the

---

[18] "On the Nature of Dreams" by C. G. Jung, from "Dreams", Bollingen Series, 2010 Edition, p. 67

[19] "The Practical use of Dream Analysis" by C. G. Jung, from "Dreams", Bollingen Series, 2010 Edition, p. 97

[20] "General Aspects of Dream Psychology" by C. G. Jung, from "Dreams", Bollingen Series, 2010 Edition, p. 46

[21] "General Aspects of Dream Psychology" by C. G. Jung, from "Dreams", Bollingen Series, 2010 Edition, p. 27

patient to describe an object that was part of the dream, but having to act like the analyst himself had never heard of the object before[22]. This would ensure a correct and accurate association of the object, free of influences from any typical description.

When the context is established, the analyst must uncover what kind of function the dream executes. Jung recognized **four primary functions** dreams can have. The first one is the **compensatory function**, in other words being a self-regulation, meaning that what the patient might be missing in real life, he experiences in his dreams. This function is useful as it demonstrates what we are longing for, since sometimes we do not even consciously realize what we desire. Applying this function to schemas would mean that in dreams we activate those schemas that we do not experience sufficiently in our waking life. The second function is the **prospective function**, where the unconscious searches for a way to prepare the dreamer for future expectations, taking a guess how the day of tomorrow could look like[23]. The **reductive function** follows, being a negative compensation, meaning that when a patient pretends to be someone better then he really is, his unconscious would display him as his true self, having a firstly negative effect on the patient[24].The fourth function is the **moral function**, teaching the patient what is right and what is wrong, in other words, the moral function searches for a way to show us to treasure our own values, and remind us what they are composed of[25]. It is very dangerous to overrate the moral function, since it is only hinting at something, not making decisions for us, nor telling us how to behave.

---

[22] "The Practical use of Dream Analysis" by C. G. Jung, from "Dreams", Bollingen Series, 2010 Edition, p. 98
[23] "General Aspects of Dream Psychology" by C. G. Jung, from "Dreams", Bollingen Series, 2010 Edition, p. 41
[24] "General Aspects of Dream Psychology" by C. G. Jung, from "Dreams", Bollingen Series, 2010 Edition, p. 43
[25] "General Aspects of Dream Psychology" by C. G. Jung, from "Dreams", Bollingen Series, 2010 Edition, p. 31

When going into detail and analyzing what each object means, Jung described two main analytical approaches: The **interpretation on a subjective level and on an objective level**. Interpreting a dream on a subjective level means recognizing that every element, such as characters and objects, can be part of the dreamers self. Interpreting on an objective level naturally means the contrary, in other words, every element can stand for itself, instead of being part of the dreamer[26]. To keep in mind is that dreams use a **figurative language** to express themselves, meaning that while on an objective level a door is not part of the dreamer, it can still stand for something else entirely, like an obstacle that must be overcome or a path that is not yet available to the dreamer[27]. Jung defended the strong belief that no interpretation can be achieved without the dreamer, meaning that the patient must be questioned and listened to, after all he is the only one who can arrive at a productive result, the analyst is merely the entity helping the patient achieve that goal[28].

A crucial mistake done by junior analysts is the tendency to prefer associations that lead to positive results, instead of those that lead to productive results. The truth is not always positive, therefore during an analysis, things must be taken as they appear to be and not as one would like them to be[29]. Another common mistake is the analyst's assumption that his psyche is similar to the one of the patient[30]. This sign of compassion leads to clouded judgment and unprofessional results. Jung also observed that as treatment progressed, dreams tend to get foggier, and more difficult to analyze, as a result of Freud's psychic

---

[26] "General Aspects of Dream Psychology" by C. G. Jung, from "Dreams", Bollingen Series, 2010 Edition, p. 58

[27] "General Aspects of Dream Psychology" by C. G. Jung, from "Dreams", Bollingen Series, 2010 Edition, p. 34

[28] "On the Nature of Dreams" by C. G. Jung, from "Dreams", Bollingen Series, 2010 Edition, p. 70

[29] "The Practical use of Dream-Analysis" by C. G. Jung, from "Dreams", Bollingen Series, 2010 Edition, p. 90

[30] "General Aspects of Dream Psychology" by C. G. Jung, from "Dreams", Bollingen Series, 2010 Edition, p. 45

entity, the censor[31]. We know that we are only able to analyze the manifest dream content, since the latent dream content runs through a series of censors before reaching our conscious memory, it is the task of the analyst to keep the content as raw and fasting as possible, leading the patient away from assumptions about certain behaviors that would only apply to the real world. When analyzing dreams the analyst is sometimes forced to reject all previous knowledge about the waking world and be completely susceptible to different rules, situations and emotions.

As a summary of Jung's theories and techniques, he stated that this was prove to demonstrate that Freud's belief of dreams being nothing more than sexual wish compensators, was naïve and completely outdated[32]. Personally I could not agree more, yet while Freud might have not established the most comprehensive dream theory, he will always be one of the founding fathers of psychology, ranging far beyond the analysis of dreams and the scope of this essay.

2.1 Jung's Archetypes and the Collective Unconscious

The core of Jungian dream analysis lies in the theory of archetypes and the collective unconscious. Jung defines the collective unconscious as a reservoir of human experiences and mythological motifs while also being the deepest level of our psyche. M5, as seen on the next page, demonstrates a simplified model of the Jungian psyche, with the collective unconscious at its deepest level. Archetypes[33] are the entities that lie in this collective unconscious, being categorized collections of human behavior, ideas, images, patterns of thought and such. Jung defined the dreams that access the collective unconscious as **archetypal**

---

[31] "The Practical use of Dream-Analysis" by C. G. Jung, from "Dreams", Bollingen Series, 2010 Edition, p. 93
[32] "General Aspects of Dream Psychology" by C. G. Jung, from "Dreams", Bollingen Series, 2010 Edition, p. 32
[33] "Archetypes", Wikipedia: "universally understood symbol, term, or pattern of behavior, a prototype upon which others are copied, patterned, or emulated.", URL: http://en.wikipedia.org/wiki/Archetype (08.04.2012)

dreams, or meaningful dreams. He observed that there are certain symbols that have made their appearance across history in many dreams as they do today, leading him to believe that there must be a realm beneath our unconscious to which the entire human race has indirect access. The collective unconscious is a part of our psychic apparatus that we inherit from previous generations; we cannot recognize its wholeness by education or other conscious efforts because it is innate[34], in other words, a priori[35].

M5 Simplified Jungian Model of the Psyche[36]

Consciousness

Ego

Personal Unconscious

Complexes

Collective Unconscious

Archetypes

M5 also demonstrates the personal unconscious where the previously discussed dreams manifest, while the archetypal dreams manifest in the collective unconscious[37]. Archetypal dreams are to be analyzed in a slightly different manner. The gathering of context is completely irrelevant as there usually is little to no context to be found. Archetypal

---

[34] "Concept of Collective Unconscious", URL: http://www.carl-jung.net/collective_unconscious.html (01.04.2012)

[35] A priori means independent from experience.

[36] Adapted from "Jungian Models of the Psyche", URL: http://eve3.wordpress.com/2008/07/06/jungs-model-of-the-psyche/ (29.03.2012)

[37] "On the Nature of Dreams" by C. G. Jung, from "Dreams", Bollingen Series, 2010 Edition, p. 77

dreams are often presented in a fairytale like, poetic form, showing behaviors and visions that have no direct connection to the dreamer. Jung states that an analyst who has no knowledge of history and mythology is unable to interpret an archetypal dream in a productive manner. The analyst is forced to draw away from the patient and refer to history and mythology to see what the presented images mean, in order to ultimately transport the findings onto the patient's life[38]. Archetypal dreams usually occur at critical stages in our life, like puberty or slightly before death and serve a very revealing purpose when treated correctly.

There is an infinite amount of archetypes within the collective unconscious; yet Jung defined **seven primary archetypes** along with their characteristics, which typically appear within archetypal dreams. **The Persona** is the image of how we represent ourselves to the world, in other words, The Persona wears our superficial values. **The Shadow**, on the other hand, is the part of us that craves to do the things we forbid ourselves to do. It is incompatible with who we think we are. The Shadow often makes its appearance as a witch or as a similar creature. **The Anima or The Animus** is the part of us that is feminine, or respectively masculine, being the opposite sex that we carry within us. It can appear to remind us that we may be too masculine or feminine. **The Wise Old Man** represents wisdom and our unconscious gifts. He is there to guide us and make us realize what we have inside of us. **The Great Mother** is the embodiment of help and love towards the dreamer, while **The Divine Child** impersonates our creativity, our fantasy's as well as our childlike happiness, being one of the most important archetypes that the dreamer usually seeks to protect. **The Trickster**, being the last of the seven primary archetypes, makes its appearance to deceive us, to lead us into an illusion, to trick us. He is not evil as he may also help us to get out of certain situations using his charm. These

---

[38] "On the Nature of Dreams" by C. G. Jung, from "Dreams", Bollingen Series, 2010 Edition, p. 78

primary archetypes are behavior patterns observed by Jung, recurring countless times in various dreams across history and mythology.[39]

Recognizing these archetypes and knowing what they express, can lead to a very fruitful analysis of an archetypal dream, and therefore undoubtedly change the patient's life. There is no exact procedure to deal with such dreams, one must research and draw connections across history and mythology to arrive at a productive result. At the end of this chapter, M6 demonstrates a Complete Framework of Jungian dream analysis, ranging from dreams manifested in the personal unconscious up to the archetypal dreams from the collective unconscious, representing the embodiment of Jungian dream theories.

2.2 Practical Dream Analysis relying on Jungian Theories

In the following I wish to present the results of a non-archetypal dream analysis. The dream is from a man whose name will not be mentioned to preserve discretion. This is the dream that he reported to me:

"My study is on fire. There doesn't seem to be a good chance to extinguish the flames as each object and eventually everything is in flames except some small spaces on the floor. Yet I am thinking how the fire could be extinguished. As I'm trying to think of a way I remember there's a party going on in the house. Then people come up to me and are trying to tell me what to do and what not to so the fire in my room can be extinguished but I don't see a way after all. They are telling me the fire has to be taken out and the situation seems to get denser. There's some sort of ditch with water at one side of the room that I notice now. So I get myself a bucket and throw some water in the flames but it doesn't seem to help. So I sit down again at the corner of the room next to the door where I've been all this time. The next thing happening is that I open my eyes and I'm lying on the floor of my room which is still on fire and I notice how each of my belongings are burning down. That is when the idea occurs to me to concentrate on extinguishing single objects to diminish the overall fire. So again I take a bucket and fill it up with water out of the ditch and

---

[39] "Jungian Dream Interpretation" by Celeste Adams with Gordon Nelson, URL: http://www.spiritofmaat.com/archive/may3/jung.htm (08.04.2012)

throw some on the object in front of me which after a while has stopped burning. Yet the object which is some sort of blanket or curtain hasn't really been damaged. There are only some holes and smut. Other objects that seem to burn for the whole time during the extinguishing process didn't fully burn down. It doesn't seem like the fire truly destroys anything. Nor does the fire spread out to any other room or influences the mood of the party which is going on one level below my room."

After having heard the dream, my first task was to get to know this person, in other words, establish the context. We started talking as he began to open up, beginning to speak about his issues, being of a mostly social nature. He felt an obligation to work, if he did not, he felt like he had so much to catch up to, like he had not done something for too long. He also explained that work was more fun for him than enjoying free time, stating that work proves his theory about life; being creative and full of innovative ideas. The only doubt he had was that if he worked too much, he might one day not like it any more. This behavior led to him being a man who was barely socially active. I asked him how he felt about that and he said that he believed that in his study he was to gain much more spending a few hours than outside around people. The only slight doubt he had about this matter was people not seeing him in the way that he wants to be seen.

With a stable context established we moved on into analyzing the dream at hand, separating it into two essential pieces; before the faint, and after he woke up on the floor, as well as choosing the **subjective analytical approach**. This moment represented a break within the dream, because afterwards everything had changed. In the beginning he is confronted with his study, the place where he spends most time and where he apparently gains the most, being torn up into flames. The flames represent the danger that lies in such belief as well as that material things cannot last forever. There is a party downstairs, representing his social distance. People come upstairs and wonder why he is not capable of extinguishing the fire, in other words, getting rid of the issue. He does not know how to handle the issue, having the solution right beside him, the ditch and a bucket. He faints and wakes up again, immediately realizing how to solve the problem, step by step. He realizes that things are not as bad as they seemed, yet the fire damaged his property if only a little bit. The party goes on and the people suddenly care little about the problem. The time before the faint is the current situation of the man, not knowing how to handle this social distance and feeling the fear of people having the wrong image of him, while the time after the faint is the possible future, demonstrating that the issue is not that hard to solve. The solution is lying right beside him, yet he does not know how to use it. The dream is serving a **prospective function**, warning the subject not to rely too much on his belief and desire to work, set against any further social interaction. While it is a lifestyle one can chose, the dream is indicating that it might not be the most suitable one for the man. From the viewpoint of causality, the dream wants to demonstrate a situation of danger, while from the viewpoint of finality; the dream intends to prevent further emotional damage.

This was one of many possible interpretations while also not being completely elaborated. A completely elaborated interpretation of a dream would go beyond the scope of this essay. While such interpretations should not be overrated, they can indeed give us insight on issues that might have not been considered before. Ultimately we make the decisions and not our unconscious, yet there is no wrong in listening to our dreams to seek better understanding.

2.3 Archetypal Dream Example

Now let us take a look at an Archetypal dream. Naturally these dreams are much longer as they are more complex. Proceeding in the same manner as with the non-archetypal dream would be unwise, because the length of an analysis of this sort of dream would go far beyond the scope of this essay. The purpose of this dream example is to get a feeling of how archetypal dreams can be. The dream was written down in a fairytale-like manner to stay true to its origins. The recognized archetypes are underlined and respectively embody their traits.

"I wasn't feeling well, I felt like I was powerless. A severe amount of energy that my body and soul depended on to fulfill what my mind needs, was lacking. The casual solution one seeks in this kind of situation is to go to the doctor, receive a magical pill of some kind and done it is. Fueled by this naive thought I made my way to the pharmacy, only to find it situated at an unexpected place. As I entered I saw the long line ranging from the middle of the store up to the reception. I stood in line for quite a while, nobody else was lining up behind me, leaving a sense of comfort in my soul. At last it was my turn, recognizing that the expected stereotype of a 'doctor' was no other than an old, very kind and sweet lady of white skin; a complete stranger; <u>The Great Mother</u>. I explained my problems to her and as the conversation moved forward I felt more and more warmth coming from this human being standing in front of me. The confidence and positive emotions that she transmitted to my mind were irreplaceable as they were indispensable. Hours passed and we started talking about my life and hers as if we had known each other for years.

We left the pharmacy together; standing on the sidewalk she suddenly asked a favor of me. She wanted me to deliver money to <u>The Trickster</u> who was in a café at the corner. The bus stop to drive home was just around that corner so I thought there's no reason to refuse. She gave me a lot of coins and I started to walk towards the café. I passed by the café, going towards my bus stop seeming as if I had forgotten the location where I should be heading. There, I was looking for <u>The Trickster</u> but I could not find him, comprehending that he was at the café I had just passed by. Feeling a little dazed I returned to the café and found <u>The Trickster</u> instantly. He was a young man, in his twenties, sharp haircut, and stylish clothing. I gave him the coins, he looked at me and said that I'm too nice to other people, emphasizing that there was no reason for me to agree to such a favor. He left the café and headed towards <u>The Great Mother</u>. I felt an undeniable urge to follow <u>The Trickster</u>. After an instant I had no choice but to summit myself to that urge. Back at the sidewalk, <u>The Great Mother</u> had a slightly different appearance, seeming even sweeter and warmer than before. She reminded me of somebody from my past life. I asked her if she was that person she reminded me of. To my disappointment and confusion, she *was not. This question triggered the urge to tell her all about this past life I had lived.*

Finding myself back at home, The Wise Old Man wanted to take me to a restaurant, in order to strengthen his relationship with me. I soon realized that this was the typical father and son activity in which I did not felt like taking part. To fulfill his desire I decided to go anyway. We were standing at the glass door getting ready to leave as he pointed through the glass, obtaining my full attention. It took me an instant to realize what was there. In front of me were many animals being more beautiful than human words would allow their description. Cats and leopards in their purest form leaving an ecstatic, indescribable emotion within me, being right there, in my stairway hall. I opened the door and started to play with them, to pet them, as if I was a little child being happier than I could have ever been before. They were so truly marvelous, of many colors and of galactic origin. Being in this whirlpool of positive emotions the whole scenery around me began to shift. The Wise Old Man disappeared and I suddenly found myself in tropical surroundings, witnessing a world being built in front of my eyes of such an incomparable attractiveness.

The animals were still there, now in their natural habitat. After a while I stopped to play with them and realized where I was; it was hot, the sun was shining and I was standing beside a hotel. There was a family who wanted to leave the hotel soon and return to their home country. As I looked to my right, I saw The Divine Child sitting a few meters away from me; she was maybe eleven years old, beautiful black hair and white skin. She was sitting on the ground speaking to a boy of equal age. I came closer only to realize that the boy was singing for her. I complimented his talents and took The Divine Child onto my back. I felt like I had to protect her somehow, like I had a connection to this stranger posing herself as divine. The Divine Child was part of the family that was soon leaving; yet the meanwhile, I spend with her on my back. We went to the center of town. We found a lot of small shops and other kids playing games on the street, with the magnificent ocean on one side and the vast rain forest on the other. Suddenly The Shadow disguising himself as a young kid punched the back of The Divine Child. I turned around as I felt an incontrollable rage, I had to protect her. I tried to hurt The Shadow but I could not, no matter what I tried nothing seemed to be effective. At last I spitted onto his head as he suddenly lights up in flames. I started to run with The Divine Child *still on my back. Suddenly the thought of my innocence came to me. I was just protecting her; he had started the conflict while I just resolved it. I stopped running as* The Wise Old Man *now disguised as a middle aged man, supposedly supervising the children's game, caught up to me. I explained the situation to him, he understood, yet was not very happy with my resolution. He told me that* The Shadow *never dies and that he would surely come back for revenge. I remembered that I had to return* The Divine Child *to her family that was soon leaving this town. As I headed back towards the hotel I saw* The Great Mother, *now being the female parent of The Divine Child, walking on the sidewalk. I gave her the girl and witnessed for my dismay that she could not remember me from before. I sensed an unquestionable need to protect this family and ensure that it got out of this town safely. I arranged a convoy to escort the family out of town, consisting of* The Wise Old Man, *two light figures of immense moral value and none other than myself. As we left town* The Shadow *made an appearance on a bicycle, slowly catching up to the convoy, the figures of light intimidated him and he did not dare to come any closer as we left him behind upon leaving the area. We passed my house and I wanted to show them where I come from, but sadly they were not interested because of what laid ahead of us: a gigantic water pool in a half frozen state in middle of tropic Caribbean surroundings. They certainly were as surprised as I was. We stopped and examined the area. I closed my eyes as I sensed my duty was fulfilled at this time, being catapulted to a world beyond my imagination; a world of magic, wizards and love.*

I rented a room in this strange town of fairytales I now found myself in. Inside the main house where the reception was located, I met The Anima, *a beautiful woman wanting to take me for a ride on a vehicle so extraterrestrial that I could not refuse. With this vehicle one could float, jump or just move with wheels. To trigger one of these modes one needed a stone of some kind.* The Anima *once possessed all three stones, yet as of now; the stone of fire to enable the wheels was missing. And so we went floating through this magical world. I could not restrain myself to ask her what had happened to that missing stone. She looked at me and planted a seed within my mind, triggering a series of pictures to appear, soon moving as fast as a film. It was the story of her life I was now witnessing, seeing before my own eyes. She was very in love, so in love that she was sure to label the love she had with another being as eternal. To create the stone of fire one must combine the element of fire with the emotion of eternal love. She stood upon a bridge having her eternal love not too far away as she summoned with her bare hands an amount of fire unimaginable to mankind right behind her. She used her entire energy to do as she did. Her eternal love was coming closer as in the midst of dangerous beauty their lips and arms reached for one another. The artistry present in front of my eyes filled my heart with joy and enchantment. True love in combination with the destructive elegance of fire created the object of desire in front of them. I felt such ecstatic happiness to witness such natural symmetry, a symmetry that was soon disrupted by one of them pushing the other away, not having noticed their success and complaining why the object of desire had not been produced. The Anima pointed at the stone laying at the side of a river surrounded by wooden statues. The fire gradually disappeared as they both stared into the stones direction. The wooden statues divinely came to life and pushed the stone into the river as they threw themselves after it. The Anima and her eternal love were frozen and could not take action as they witnessed the hundreds of wooden statues coming to life and diving after the object of desire following the flow of the river, being lost forever at the mercy of paranormal forces. The conclusive scene left no more than the eternal love of two individuals to be in the eternal space and time of that mythical world that once was."*

One can easily recognize the poetic manner in which the dream presents itself. The different scenarios are only connected by the various Archetypes disguising themselves differently each time, yet embodying the same traits, directing the dreamer's attention towards a certain matter. An example would be the couple in the last scene, using their love, combining it with fire only to produce a materialistic object. Yes, their love was eternal, yet they still let their ambition come in the way, resulting in the loss of that precious object. That was just a simple presumption of what the anima was trying to transmit to the dreamer, yet one should go deeper, relying on history and mythology to analyze the motives of ambition, materialism, fire and love for this particular scene of the Archetypal dream. These dreams are unlike any other dreams, sending us on an unforgettable journey of experience, beauty and self-exploration.

M6 Complete Framework of Jungian Dream Analysis[40]

Personal Unconscious			Collective Unconscious	Archetypes	
	Analytical Approaches	Objective Level			The Trickster
		Subjective Level			The Divine Child
	Dream Function	Moral			The Great Mother
		Reductive			The Wise Old Man
		Prospective			The Anima/Animus
		Com-pensatory			
	Twofold Viewpoint	Causality			The Shadow
		Finality			The Persona

---

[40] Adapted from "Dreams" by C. G. Jung, Bollingen Series, 2010 Edition

3. Stephen LaBerge and Lucid Dreaming

Steven LaBerge is a psychologist who began the studies for his PH.D[41] at the university of Stanford, on the subject of lucid dreaming, which he received in 1980. He is a pioneer in the field of lucid dreaming and the founder of the Lucidity Institute, an organization devoted to the scientific research of lucid dreams[42]. He also proved that dreaming occurs in the REM stage. He discovered that the eye movements in a dream are actually happening physically as well. Meaning that when a dreamer looks to the right in a dream, his eyes are actually looking to the right in the real world, underneath the closed eye lid. In the lab, dreamers agreed to move their eyes in a specific pattern when they were lucid within a dream, establishing a connection between conscious dreamer and conscious scientist and ultimately proving that dreaming occurs in the REM stage.[43]

That leads us to the question of what a lucid dream is. Lucid dreams are dreams where the dreamer is fully conscious, being aware that he is dreaming[44]. This behavior enables the dreamer to have full control over his dreams and live adventures far beyond this world, leaving an indispensable feeling of freedom behind. Old inscriptions show that lucid dreaming has been known for centuries. Lucid dreams can appear spontaneously in people's life and they can be invoked by training. Everybody can learn to lucid dream if enough effort is put into the task at hand. In the following I wish to demonstrate which techniques LaBerge developed to enable frequent lucid dreams, as well as take a look at the advantages and disadvantages of such ability.

---

[41] Doctor of Philosophy
[42] "Stephen LaBerge", Wikipedia, URL:
http://en.wikipedia.org/wiki/Stephen_Laberge (08.04.2012)
[43] "Exploring the World of Lucid Dreaming" by Stephen LaBerge and Howard Rheingold, Random House publisher, 1991 Edition, Chapter 2, p. 23-24
[44] "Exploring the World of Lucid Dreaming" by Stephen LaBerge and Howard Rheingold, Random House publisher, 1991 Edition, Chapter 1, p. 11

3.1 The Role of Awareness

Awareness is crucial in the process of learning how to lucid dream. We say we are aware all the time yet one would be surprised to know how little some people notice about the world around them. Being aware means to notice things, to pay attention to ones surroundings and frequently look for odd things that just do not correlate with the world one knows. The theory is as follows; if one is constantly aware in the real world, looking out for dreamlike things, then when one is really within a dream and doing the same, the result will be positive, meaning that developing a critical attitude towards the world can enable you to lucid dream as well as notice things that you never noticed before[45] . It is evident why this works, since our attitude in the real world is known to carry over to our dreams. This leads us to **ADA**, namely 'All-Day-Awareness'[46]. This is a method many people use to effectively lucid dream. The meaning of ADA is self-explanatory; educate yourself to be constantly aware. Ironically there are those who have ADA without even knowing about it, for them it is natural, it is a talent. For others it is hard work until ADA happens automatically. Awareness is essential to achieve frequent lucid dreams, leading us to critical state testing and dreamsigns.

3.2 Critical State Testing and Dreamsigns

Performing a **critical state test** means to question your reality and seriously test if you are dreaming. This might be harder than it sounds, since one easily answers this question with 'no', before even critically testing ones reality. Yet that is the key, since a frequent habit of per-

---

[45] "Träum' ich oder Wach' Ich?" by Paul Tholey and Kaleb Utecht, from "Schöpferisch Träumen", publisher: Klotz, fifth edition 2008, p 49-50
[46] "ADA" by Chewnie91, URL: http://www.dreamviews.com/f12/ada-all-day-awareness-dild-129237/ (08.04.2012)

forming critical state tests crosses over into the dream world, delivering many lucid dreams.[47]

Another crucial step towards the goal is realizing what is dreamlike about our dreams, in other words, detecting **dreamsigns**. If one knows what is dreamlike, one knows what to look out for. Dreamsigns are usually things that cannot be real, things that can only occur in the dream world[48]. LaBerge defined **four categories** dream signs can belong to, with **inner awareness** being the first category. Things that happen within us, like unusual emotions or thoughts, belong to this category. Actions or behavior patterns of the dreamer or of other dream characters that are unusual belong to the **action** dreamsign category. If something or someone suddenly changes its shape, the dreamsign belongs to the **form** category. The last category is the **context** category, with unusual situations or places that make little sense belonging to it[49]. The categorization of dreamsigns can help to recognize the most recurring kind of dreamsign, in order to set the target on the respective category. LaBerge called the categorization of dreamsigns using his system the **dreamsign-inventory**. Recognizing what is dreamlike in our dreams can easily trigger a lucid dream.

3.3 Lucid Dreaming Methods

LaBerge developed many lucid dreaming methods to induce them almost at will. These methods differ from the previously stated ones as they are to be executed in the moment of desire of a lucid dream, instead of frequently during the day. A crucial prerequisite to be successful with any of the mentioned methods is to believe in success. If one does not believe he or she will be successful, there is little to no chance of success. On the other hand if one firmly believes in success, using

---

[47] "Exploring the World of Lucid Dreaming" by Stephen LaBerge and Howard Rheingold, Random House publisher, 1991 Edition, Chapter 3, p. 59
[48] "Exploring the World of Lucid Dreaming" by Stephen LaBerge and Howard Rheingold, Random House publisher, 1991 Edition, Chapter 2, p. 41
[49] "Exploring the World of Lucid Dreaming" by Stephen LaBerge and Howard Rheingold, Random House publisher, 1991 Edition, Chapter 3, p. 43-46

auto suggestions to stabilize that belief, chances of success are relatively higher[50]. Good dream recall is also essential, yet this is a skill that comes along naturally when deciding to spend more time with our dreams. Keeping a dream journal is the first step towards good dream recall. When dreams are written down many more details can appear along the way, and therefore enrich the experience.

3.3.1 DILD Dream-Initiated-Lucid-Dream

This is not a method, rather the scientific term for a spontaneous lucid dream. The dreamer realizes during his dream that he is dreaming, by recognizing a dreamsign, performing a critical state test, or by pure luck. Studies show that this is the kind of lucid dream that is most likely to happen among dreamers[51]. People with poor dream recall probably have had DILD's that they just cannot remember.

3.3.2 MILD Mnemonic-Initiated-Lucid-Dream

This is a method specifically developed by LaBerge, which helped him achieve lucid dreams completely at will. The method relies on being able to remember to do things in the future, recognizing that memory does not only work backwards, but also forwards. Therefore, to execute this method successfully the foremost prerequisite is a good prospective memory. One can attain this by practicing to remember things in the future while being awake. When one has a good prospective memory, the execution of the technique is simple. Before going to bed the dreamer must set the clear intention of remembering that he or she will soon be dreaming. When in a dream, the memory should appear and remind the dreamer, triggering a lucid dream[52].

---

[50] "Exploring the World of Lucid Dreaming" by Stephen LaBerge and Howard Rheingold, Random House publisher, 1991 Edition, Chapter 3, p. 79-81
[51] "Exploring the World of Lucid Dreaming" by Stephen LaBerge and Howard Rheingold, Random House publisher, 1991 Edition, Chapter 4, p. 95
[52] "Exploring the World of Lucid Dreaming" by Stephen LaBerge and Howard Rheingold, Random House publisher, 1991 Edition, Chapter 3, p. 73-75

3.3.3 WILD Wake-Initiated-Lucid-Dream

This is by far the most interesting, difficult and effective method, consisting of falling asleep consciously. Mastering this method is very complex as it is the kind of lucid dream that is less likely to happen. The technique consists in migrating over from the real world into the dream world while tricking the body into thinking we are asleep but actually being wide awake during the process. For this to work one must take into account the sleep cycles discussed earlier. Falling asleep consciously means to cross over into the REM phase with a mind that is awake and a body that is asleep. If one would try this technique at the beginning of the night, with no previous sleep, one would have relatively little chance of success because of the N4 sleep before the actual REM phase, which would anyhow be very short in the first sleep cycle. To execute this method with success, it must be performed after at least four hours of sleep, in other words, after two and a half sleep cycles, right before the next, now longer lasting REM phase.

So how do we fall asleep consciously? The key is to trick the body into thinking we have gone to sleep by not moving at all and staying very relaxed, yet aware. It is very easy to wander off and fall asleep during this process as it would be the natural outcome of such a state. If successful, one will begin to see hypnagogic imagery, also called **HP**. HP can present itself in light bolds of many colors flashing and crossing over each other, eventually creating a dream scene in front of our closed eyes. The cue to stay awake during this process is to focus on HP and keep the mind active while the body falls asleep[53]. Right before entering the REM phase, the body goes into sleep paralysis, if mentally aware at this state, the dreamer will experience a wave of vibrations going through his entire body, eventually going numb. This scares many beginners, yet it is the most natural thing to happen, the only difference is that we are usually asleep in N2 or N3 at this stage. When

---

[53] "Exploring the World of Lucid Dreaming" by Stephen LaBerge and Howard Rheingold, Random House publisher, 1991 Edition, Chapter 4, p. 96

the HP starts to evolve into the dream, the conscious dreamer finds himself as a spectator and witness to the creation of a whole scenario. Once the dream is created and stabilized, the dreamer finds himself in a lucid dream, more conscious and aware than in any other kind of lucid dream. The dream will seem very realistic, almost completely indifferent from walking reality[54].

The reason why this technique is essentially difficult to master lies in the procedure itself. It is so easy to react to a twitch, or to an itch, yet to succeed one must withstand the urge. The body wants to test if we are really asleep by sending these impulses, it is our task to fool it into thinking we are unconscious.

3.4 Advantages and Disadvantages of Lucid Dreaming

Every coin has two sides as does lucid dreaming. In the following I wish to expose advantages and disadvantages of such ability.

So what can lucid dreaming do for you? It can change your life; give you an insight into your true self. When we are fully conscious within the dream world we have the power to control unconscious creations, in other words, we can control the dream. This power enables us to create any scenario we desire. We can speak to people and see what they say, and above all, we can speak to ourselves, to our unconscious. This can have a very positive effect, if we listen to what we have to say to ourselves. In lucid dreams we can practice for the day to come. Maybe we have a big speech coming up that we would like to rehearse. We have a world, a space that we can shape to our likings, using it to learn, create beautiful music, and to let our fantasies go wild. We can play out scenarios and observe how certain characters behave to certain questions. And above all, we can be free, flying to the end of the Milky Way and back.

---

[54] "Exploring the World of Lucid Dreaming" by Stephen LaBerge and Howard Rheingold, Random House publisher, 1991 Edition, Chapter 4, p. 94-99

Such power comes with a great responsibility and can easily be abused. We must remind ourselves that the realm of dreams is not the real world and that all characters are our creation and not real human beings. One can easily want to spend his entire life in a lucid dream, neglecting friends and family. Manipulation is not the goal of becoming lucid[55], we must remember to stay true to who we are and to who we want to become. C. G. Jung stated that no man can control a dream meaning that dream control was divine and impossible to achieve, he was strictly against lucid dreaming[56]. It is true that lucid dreams cannot be analyzed in the same manner as normal dreams, yet it is foolish to believe that they cannot be analyzed at all. One can easily let go and let the dream happen, while being able to analyze during the dream itself, giving us a deeper insight into our very own mind[57]. The power of dream control has to be handled with responsibility, in order to have unforgettable experiences that can shape who we are in a positive way.

Lucid dreaming is an incredible experience that can influence us greatly, yet we make the decision if the influence is a negative or a positive one. Responsibility over our actions in both worlds, while enjoying the things that each world gives us, is the key to ultimate success as a dreamer as well as a human being.

Naturally this chapter merely served as an introduction to the world of lucid dreaming. If there is an interest in this topic I sincerely recommend LaBerge's book *"Exploring the World of Lucid Dreaming"*. This book serves as an extensive and complete framework which will lead the reader to the ability of lucid dreaming.

---

[55] "Exploring the World of Lucid Dreaming" by Stephen LaBerge and Howard Rheingold, Random House publisher, 1991 Edition, Chapter 6

[56] "Das große Praxisbuch der Traumdeutung" by Klausbernd Vollmar, publisher: Knaur, 2011 edition, p. 24-25

[57] "Exploring the World of Lucid Dreaming" by Stephen LaBerge and Howard Rheingold, Random House publisher, 1991 Edition, Chapter 2, p. 31

4. Philosophical Aspects

In the last part of this essay, I wish to go into some philosophical aspects in relation to the discussed topics. I believe that there is more to dreams then we can imagine. I also believe that the connection between the two worlds is far greater than most people describe it. With all the previously gained knowledge, it is time to think about what dreams are and what connections lie between people like LaBerge and C. G. Jung. Reality and how we recognize it stands in the middle of the following, where I wish to illustrate that dreams are far more than most people give them credit for.

4.1 Similarities between LaBerge, Plato and Jung

The title of this heading might sound confusing as there appears to be no obvious similarity between their theories, yet when taking a closer look, they are all aiming for the same; explaining how we are able to recognize objects as we do, if in the dream world, or in our world.

Plato was a Greek philosopher who lived around 400BC. He was the student of Socrates and is mostly know for laying the philosophical foundations in the western world, as well as for his theory of forms, also known as theory of ideas. He states that there is a space called the realm of forms where all human ideas are stored[58].

He defines an idea, just like LaBerge describes a schema; a model of something, containing all the necessary traits for us to recognize it. The difference is that LaBerge never explained where these schemas come from; he only said that they are stored within our mind. Plato states that all ideas are within us, a priori. Before we enter this world, we go through the realm of ideas where all ideas are sealed within us, making the task of our life to recognize these ideas. For LaBerge we create schemas every day, combining old ones and discovering new things, yet it is very relatable, Plato only expanded what LaBerge said later, ex-

---

[58] "Plato", Wikipedia, URL: http://en.wikipedia.org/wiki/Plato (08.04.2012)

plaining with his realm of ideas where these schemas or ideas were first originated. So where does Jung fit into this? While Plato and LaBerge, with their ideas or schemas, try to explain how walking perception works, Jung, with his collective unconscious and archetypes, tries to explain how perception works within the dream world. The difference is thinner than it may appear. An archetype is a collection of human thoughts, behavior patterns as well as images, forming something concrete to be recognized by us in the dream world, just like ideas and schemas form something concrete to be recognized by us in the walking world. The collective unconscious and the realm of ideas are amazingly similar, both being the realm where these concrete ideas of things are stored, being within us a priory and indirectly accessible to the entire human race. M7, seen below, demonstrates the three theories in relation, simplifying the similarities between these three thinkers for a better overall understanding. These similarities demonstrate how great thinkers like Jung and LaBerge were building on Plato's theories in their own way. It also shows that we are all just trying to explain the world, explain how we can perceive and recognize the way we do. Fact must be, that there is some kind of space where categorized knowledge is stored for us, allowing us to recognize in the way we do, be it named the collective unconscious or the realm of ideas, be it archetypes, schemas or forms, in the end those are just different names to explain an undeniably similar behavior.

M7 Similarities between Plato, LaBerge and Jung

LaBerge	Plato	Jung
Human Brain	Realm of Ideas/Forms	Collective Unconscious
Schemas	Ideas/Forms	Archetypes

4.2 René Descartes in Relation to Lucid Dreaming

René Descartes was a French philosopher who lived in the seventeenth century, best known for his statement, "I think therefore I am"[59]. One of his maybe not so famous thesis is the **dream argument**. The dream argument consists of the assumption that there is no way to be sure of our current state, if we are awake or if we are dreaming. Descartes recognizes that our senses have deceived us many times, concluding that one should never trust those who have deceived us once. This leads Descartes to the resolution to call into doubt everything he has known so far and build new foundations to begin with[60]. He eventually did so, publishing the results as one of his most famous literal pieces; "Meditations on first Philosophy"[61]. This procedure forced Descartes to acquire an attitude of an extremely critical nature towards the world, an attitude we referred to as ADA. This completely wild connection might not seem justified, yet when examining both attitudes it becomes clear that Descartes was already training himself to lucid dream, even if consciously unaware of that fact. By demolishing the foundations, in other words, the culture that teaches us which world is the reality and which world is not, he would eventually become a new man, able to differentiate the dream state from reality.

The dream argument is not accepted by modern lucid dreamers, as for them there are clear signs to distinguish between the two worlds, yet without proper training they are barely able to make that distinction. In dreams the maddest things can occur and most times we still firmly believe that it is our reality. While there might be clear differences, we are not able to recognize them without proper training.

---

[59] First mentioned in "Discourse on Method" by René Descartes, 1637

[60] "First Meditation" by René Descartes, from "Meditations on first philosophy", Cambridge University Press, 1996 Edition, p. 12-15

[61] "Meditations on First Philosophy", Wikipedia: "The book is made up of six meditations, in which Descartes first discards all belief in things which are not absolutely certain, and then tries to establish what can be known for sure.", URL: http://en.wikipedia.org/wiki/Meditations_on_First_Philosophy#cite_note-0 (08.04.2012)

4.3 Dreams and Reality

"Dreams are real while they last. Can we say more of life?"

– Henry Havelock Ellis

At last I wish to discuss the above quotation from Henry Havelock Ellis[62]. Dreams are a beautiful thing, a world different to this one, a world that is our unconscious creation. Yet what value does that world hold, what should it mean to us? Life is short, and sadly we live in a world full of hate and lies, leaving us no choice but to love and fill our heart with optimism and joy, because life itself is the most beautiful thing that we should enjoy despite all the negativities around us. We are human, and humans have an undeniable hunger for knowledge, a hunger that drives some of us to do terrible things, just to obtain our goals in the end. Yet we act like this is the only world that matters, we act like there cannot be anything else, like understanding is everything. I believe that such picture of the world is far too narrow. Beyond the world we call reality lies so much more. I am aware that there is no way to prove such wild assumptions, yet prove is for humans who refuse believing, who refuse to look at the worlds beyond this one, like the realm of dreams.

Our culture teaches us to ignore dreams and to focus our entire existence onto this reality, without opening the mind to things that lie beyond this plane. Yet we make our own choices, and we can choose to open up, we can choose to love as well as we can choose to believe. A narrow view of the world is going to limit us in our experience. Gathering knowledge is an undeniably necessary process, as it is a beautiful one. When combined with love, compassion, belief and an open mind, it can be rewarding as it can shape us to a new being, it can shape us to

---

[62] "Henry Havelock Ellis", Wikipedia: "Henry Havelock Ellis, known as Havelock Ellis (2 February 1859 – 8 July 1939), was a British physician and psychologist, writer, and social reformer who studied human sexuality", URL:
http://en.wikipedia.org/wiki/Havelock_Ellis (08.04.2012)

who we really want to be. The world of dreams is a new world for us to explore, for us to experience. Humans should try their best to enjoy every experience to the limit, with an attitude as such; life can be nothing but joy. Live every second of your life to the fullest, no matter in which world you are. Reality is nothing but a term to justify our actions in this world, a dream is just as real as life is while it lasts, so why not make the most of it? We do not know what will happen after this dream of life comes to an end, yet with the mentioned attitude, dreams, life, reality, whatever you wish to call it, can be a marvelous and truly ecstatic experience.

Conclusion

"Where love rules, there is no will to power; and where power predominates, there love is lacking. The one is the shadow of the other."

- C. G. Jung

This is the end of the first essay. I have built a solid structure of knowledge on the subject of dreams, including many topics. The goal was to demonstrate the possibilities that the dream world offers, from being a Jungian analyst to being a lucid dreamer. Each of these topics could indeed cover many books on their own, yet my aim was to provide the general knowledge along with some details.

This newly acquired knowledge is advised to be used from now on, one can start with analyzing one's own dreams and see how he or she profits from it. If there is interest in a particular topic discussed in this essay, I recommend reading the books seen in the bibliography at the end. Those were the books I used to obtain information about the various topics, combining it with my own.

In the end it is up to us what choices we make, and what we believe in. As long as we believe in something, we are assured to enjoy our lives.

Bibliography

Descartes, Rene. *Meditations on First Philosophy*. 2011 Edition. Edited by John Cottingham. USA: Cambridge University Press, 1996.

Jung, Carl Gustav. *Dreams*. 2010 Edition. Translated by R.F.C. Hull. USA: Bollingen Series, 1974.

LaBerge, Stephen. *Lucid Dreaming*. Canada: Sounds True, 2009.

LaBerge, Stephen, and Howard Rheingold. *Exploring the World of Lucid Dreaming*. Mass market Edition. USA: Random House, 1991.

Tholey, Paul, und Kaleb Utecht. *Schöpferisch Träumen*. fifth Edition 2008. Germany: Klotz, 1995.

Vollmar, Klausbernd. *Das große Praxisbuch der Traumdeutung*. first Edition. Germany: Knaur, 2011.

Dream Views. no date available. http://www.dreamviews.com.

Brain, Marshall. *HowStuffWorks*. 1998. http://www.howstuffworks.com.

Foundation, Wikimedia. *Wikipedia*. 15. January 2001. http://www.wikipedia.org/.

Garrison, Cal. *The Spirit of Ma'at*. 2004. http://www.spiritofmaat.com.

Swierzewski, Stanley J. *Health Communities*. 1998. http://www.healthcommunities.com/.

The New York Times Company. *Psychology About*. 1995. http://psychology.about.com/.

The Romanian Association for Psychoanalysis Promotion. *Carl Jung Resources*. no date available. http://www.carl-jung.net/.

The Third Eve. no date available. http://eve3.wordpress.com.

Thiruvelan. *Healthy-ojas*. 2011. http://healthy-ojas.com.

Part II

Subjective Realities from Within

A Journey to the Edge of Consciousness

Prologue

Dreams are an extraordinary brain twister. Nobody really knows where they come from, nor do they know how thick their link to reality really is. Personally, I have always found dreams a fascinating and very interesting subject to talk and think about; however my thoughts have not gotten me any closer to the answers of my questions, since nobody has the wisdom to respond to them. I usually remember my dreams, most of the time it is nothing special, while in other occasions it indeed is something special, but finally I am always left unsatisfied when it comes to interpreting their connection to our world. I have read that the chance to remember ones dreams increases dramatically if one spends more time with the subject at hand, after all, it makes sense, if I am truly interested in a certain matter I will always remember more clearly then when I do not care about the matter at all. In college we barely talk about the subject and if we do, we never look at dreams with the profundity that I would expect. I know many people that do not really care about their dreams at all; my father is one of them. He basically ignores them completely, in other words, he does not give them any meaning at all. Sometimes he even talks to my mother about how crazy I sound when I tell them one of my dreams at the dinner table. For him dreams and reality have no connection at all and because of that, dreams are being completely ignored by him.

It was already late when I was brushing my teeth, while staring at myself in the huge mirror which covered the entire wall in front of me. I left the bathroom thinking about how wild and crazy some dreams can get and how enormous the variety of emotions presented in dreams is. When I reached my room the first thing that spotted my eye was the beautiful view one could have when lying in my bed, observing the majestic full moon and the stunning stars in the cloudless black sky through my roof window. I slipped into my warm bed and under the soft, fresh blanked. While lying in bed I started to think about the past, my only sister had died five years ago, we all learned to deal with her death and to move on, except my father, it was then when he began reacting to surreal things in the way that he does now. It almost seemed like he never truly moved on, like he is still living in the reality that once was, yet here I was staring at the pitch black sky filled with the beautiful things of this world.

I could slowly feel how my muscles relaxed and how my breathing began to follow a specific pattern. My eyes closed, followed by the remaining sense of the consciousness leaving my body and finally, of the unconscious claiming the control over my mind, taking me beyond this world, in to a world where everything was possible, into a world over which my consciousness had no control, a world which exceeds my wildest fantasies: *the realm of dreams.*

Chapter 1

I slowly regained consciousness, hearing nothing but the growling thunder coming from outside. After I opened my eyes, I realized that it was just past midnight and that I had barley slept at all. These storms just kept getting worse and worse, claiming the lives of so many almost every night, causing such mass destruction. Is he angry with us or is he just sending us a warning? I spend most of my nights restless, just like this one, desperately longing for the solution but finally always being unable to achieve anything useful, even I felt completely powerless against them. Looking out of the windows of my fortress into the pitch black night filled with furious water drops motioned by the demented wind, I realized that the people of Pangaea are not to suffer any longer, concluding that I had to respond to these catastrophes. I decided to hold a press conference at noon tomorrow, letting the people across Pangaea and the Panthalassic Ocean know that I, the Emperor, am not going to stand by and just watch. I spend the rest of the night writing the speech and thinking about how much time had passed since my inheritance to the throne. My father was a great man while being a horrible Emperor; he was just to kind, too sweet and too weak. Back then I made it my mission to succeed where he failed, to unite Pangaea and conquer the Panthalassic Ocean. I did ultimately achieve that goal only to realize that my career as the Emperor had only just begun.

It was almost sunrise when the storm finally passed away. I kneed down on the floor and thanked him for stopping what he had started. We have been very grateful to him, to Susanoo the god of the storms and seas, but yet he makes us suffer almost on a daily basis taking just too many lives from this world. Suddenly someone knocked on the door, it was my personal Advisor, serving my breakfast.

"Good morning my Emperor, you look awfully tiered," he said in a negative tone.

"Good morning, yes I'm fully aware of that, I have a request: please arrange a press conference to be held at noon immediately," I said rapidly.

"I see you haven't been sleeping at all again?" As always, he was very worried and concerned with me.

"How could I, with my empire suffering from such a mass destruction every night!" I replied rather upset.

"I apologize, it will be arranged immediately," he answered cautiously, leaving my breakfast on the table.

After breakfasting I went for a walk at the beach to prepare myself mentally for the press conference. It was never easy, I have always felt allot of pressure and obligation, which led to myself being mentally not always too stable. In my time as the Emperor I had turned this world into such a beautiful place to be, always having the gods to watch our back, but after the storms started a couple of weeks ago that warm feeling was slowly being replaced by the impression of the gods, especially Susanoo, having turned their back on us. In the glowing sky above me the gigantic full moon was just leaving, making the sunlight of the day space to be. It was all the work of the gods, of the siblings Tsukuyomi, god of the moonlight and Amaterasu, god of the sunlight. They had a horrifying fight in the days of creation, leading to them being unable to stand each other's company. This explains why one never sees the sun and the moon fully together; it is either day or it is night. Tsukuyomi, Amaterasu and Susanoo are our gods, our protectors, our greatest fathers and mothers, yet one of them is positioning himself against us, against his own creation. We can do nothing but pray, I thought, since sacrifices were being taken enough by every storm. I was relieved but also nervous when I found the vehicle ready to go at my return from the beach. I went upstairs and got properly dressed. At my arrival downstairs, I encountered a great change of circumstances.

Chapter 2

"My Emperor, I apologize but the press conference will have to be postponed, you have to be escorted to the TRI immediately, they have something to show you," my Advisor stated in a worried tone mixed with a slight essence of happiness.

"What is this about?" I answered sort of annoyed by this change of circumstances.

"I do not know, it has been classified as top secret," he said in a wondrous tone.

"I understand, postpone the conference to the afternoon and take me to the TRI immediately."

I was not sure what to think at this point, the TRI, short for Technological Research Institute, was the only and therefore biggest research institute of the continent, with the task of bringing us new innovative technologies every now and then, what could they have so important to show me that could not wait any longer? With that thought in mind, I entered the vehicle and started leaving the area. The ride to the TRI was emotionally devastating for me, witnessing the damage from the last storm. It was a beautiful day, the sun was shining and the sky was glowing. I felt such controversial emotions inside of me, the mass destruction across the landscape slowly melting with the beauty of this world.

At my arrival at the TRI I was well received by the General of the institute, opening the door of my vehicle.

"Welcome my Emperor, we have got brilliant news for you, would you be so kind to follow me into the building?" The General said in a very formal and respectful tone.

"Greetings General, this better be good," I responded with a smile on my face, stepping out of the vehicle and admiring the incredible architecture of the building.

It was a huge skyscraper, one of the tallest buildings on the continent build mainly of glass, with the sunrays shining through it from behind. Such a beautiful sight, I though while realizing that I could not recall the last time I had been here.

"Everything beyond this point is classified as top secret, in other words, no one but you and I beyond this point," the General said in a dark, more serious voice, glaring at my curious Advisor who was standing behind me.

We were standing in front of a huge metal door, which left the impression of a very safe place within me. The door opened, the General and I, slightly after one another, entered the room. It was a huge lab with several people working, mainly sitting in front of machines and computers.

"For the past 24 hours me and my team have been developing a solution to our current situation, we have developed a force field which will soon be spawned across the entire continent at night, protecting it from everything, including the storms," the General said in a very lectures tone, "would you be so kind to take a look at the screen my Emperor?" He added.

The screen in front of us, covering almost the entire wall, presented a simulation of a storm developing in Pangaea, before it could cause any harm, it was pushed out of the continent onto the Panthalassic Ocean by the force field, the sight of such an invention was truly impressive.

"General, I am impressed, this is a magnificent invention," yet inside of me, something was bothering me. "How long will it take to install the force field around the continent?" I added.

"It should not take too long, we can have the major areas covered by tonight," the General replied.

Perfect, I thought, I shall announce this as solution to the situation at the press conference. "Thank you General, but I have to go now, I have some urgent matters to attend to," during those words, it was slowly becoming clear to me what had been bothering me about this.

"Of course my Emperor, let me escort you outside" the General replied.

"You do realize that by making this move, we would be setting ourselves in a position against Susanoo and probably against all the gods?" I was worried, this seemed like a great solution, but something just was not right, because ultimately it would be against Susanoo's will.

"With all due respect, my Emperor, we don't have a choice, Susanoo won't listen to our prayers, what are we supposed to do? We need to take action, human action to protect this world and its people!"

The General was right, but yet something continued to bother me. "General, what if these storms are only a warning of something far worse to come, what if the gods just don't find another way to communicate with us?" I responded.

"It is not our task to decrypt messages from the gods, but to save this world and as many lives as we can," he argued.

In that very moment I realized that the position against the gods, would be one worth taking to fulfill my duty as the Emperor. Yet the situation was very delicate, I thought while entering the vehicle, presenting the force field at the press conference would indeed be a very smart move, on the other hand, if this takes the wrong direction, if I was indeed right about the storms being a warning, then in my power there would be nothing left to do to save this world. A war against the gods is certainly a lost one for us humans.

Chapter 3

The sun was leaving, the moon already visible upon the brilliant dark blue sky at my time of return to the fortress. The press conference was exhausting and did not go too well. This world is a world of believers, the people trust their gods more than they trust me. Too many questions leading to catastrophes or even to the apocalypse were asked. To many question to which I just could not respond. Ultimately everybody, including myself were afraid of the worst to come, of a situation that even I, the Emperor, could not handle, of a situation that could very well mean the end of this world and my failure to protect it, I thought while staring into the dark night that was slowly manifesting itself in front of my eyes.

"My Emperor, the General called and confirmed the initialization of the force field around most areas of the continent, including our area of course, in other words, most of Pangaea will not suffer tonight and then tomorrow, if everything goes well the entire continent will be protected," my Advisor, suddenly standing behind me, said in a much relived tone, naïve as he was.

"I see, thank you, now please leave me," I had no reason to be relieved at all, on the contrary, I was petrified and anxious to know how this night would turn out to be, above all the question of how the gods would react to the force field arose, because ultimately, this night could very well be the end for us all.

The hours past and nothing happened, no storms until shortly before midnight, furious water drops started to hit the windows of my fortress, which meant that it would not be long until the storms developed themselves. Emotions never before experienced started to appear inside of me. I slowly began regretting my choices, before anything bad even happened, I just was not sure of anything anymore. My duty it was to protect this world and its people, even if finally this meant doing something against the gods will.

My thoughts were interrupted by the ground shaking and my glass falling from my table, as well as several other items, I could feel my heart beat quickening, while slowly realizing that it was a minor earthquake that soon stopped. This gave me an even worse feeling which got the opportunity to expand itself in my mind and soul, as several storms started to evolve outside of my fortress, quickly being pushed outside of the save zone by the force field onto the Panthalassic Ocean. It seemed like everything was working well.

My door harshly opened, it was my Advisor, looking pale and devastated, "Emperor, you have to come with me immediately, the General called, we need to go to the TRI!"

"What? What is happening?" I responded rather afraid.

"I do not know, but he said it is more serious than ever, he said that we have a situation to which we need a solution immediately," my Advisor insisted, he was truly afraid.

I had to hold myself together; I had to be strong, now more than ever. The force field reached from my fortress to the TRI, which meant that it was harmless for us to get there.

At my arrival we rushed into the building and saw the General in a similar emotional state of devastation.

"Emperor, please take a look at the screen, the recent earthquake was no regular one, in fact it is the first in a series of earthquakes which will ultimately split the continent into three pieces!"

The images on the screen were negatively overwhelming, I was petrified in that moment, realizing that this might be of what the gods were warning us, but I had to be strong and show no signs of weakness.

"General, what are our options?" I asked.

"We have no options, all we can do is pray to the gods, this is beyond anything we could ever resolve," the General replied desperately.

"Pray you say, pray to the very gods their messages you did not care to interpret?" I stated rather upset.

"Emperor, forgive me if I made a mistake —"

"Stop it right there, the mistake is mine as it is yours," I rapidly interrupted.

"Thank you my —" He was again interrupted by another earthquake, this time bringing us down to our knees.

"So this is how it's going to be, ironic how the gods bring us to our knees by their means," I said in an emotionally unstable tone.

The faces of the scientist with kids and family were annihilative to witness. We all knew that our chances to die tonight were higher than ever, we failed, I failed.

Suddenly the melancholic atmosphere of the room was interrupted by several incredibly loud explosions coming from outside. It did not take us long to notice that the storms had broken through the force field leaving nothing but their ashes behind. He was angry, he had to be. But finally I asked myself what was truly happening here, was this Susanoo fighting against nature, while him actually being part of nature itself? So was this the gods fighting against themselves, against their own ego? Or was nature in this case simply a destructive force on its own.

Looking around me, the ground started shaking again, everybody was doing nothing more but praying. At this point I had no other choice but to submit myself to the gods and pray as well, since my power over the situation had vanished long ago.

Chapter 4

We had nowhere to go, with the storms outside slowly taking the building apart, and the ground shaking stronger and stronger, causing some of the machines in the room to explode.

"Emperor, are we going to die here?" my Advisor asked while looking deep into my eyes and soul, sharing his horrific and petrified emotions with me.

I had no answer for him; in fact nobody had an answer to that question. The death resembling atmosphere of the apocalyptic scene was yet again interrupted by one of the remaining machines making a loud, constant, very high noise, implicating that the situation could indeed get worse. The General stood up and dragged himself to the screen. His eyes opened widely and his pupils suddenly seemed undersized. His facial expression as a whole resembled the realization of death itself, standing right in front of him.

"This is impossible. . . absolutely impossible! The satellites outside our atmosphere have spotted a black hole manifesting itself right in front of our planet, being bigger than the planet itself and therefore threatening to devour our entire world!"

My Advisor, still expecting the answer to his question, collapsed behind me. I was not able to say a single word, as the remaining machines in the room exploded and the walls slowly began falling apart. This was indeed the end; this was my failure to fulfill my duty. The storms slowly passed away, leaving only the water drops to be. Susanoo, was this your true will? Is this what you wanted for us? And what about the others, Amaterasu and Tsukuyomi, are you just going to stand by and watch this world be destroyed and devoured by death itself? Is it not your task to protect this world, when I cannot anymore?

We slowly dragged ourselves outside, having the once beautiful building crashing behind our backs, the ground shaking, dividing itself and the water drops falling upon our skin giving us a sense of coolness, a sense which was quickly taken from us by the apocalyptic scene manifesting in front of us. The sky turning from a very dark blue into pitch black, the clouds being pulled into the darkness and the water drops changing their direction, rising up into the end, leaving nothing behind but cold air, as the black hole opened in front of our eyes resembling our reality melting with another. Inside the hole was nothing to see but the end of our path, nothing to see but the obscurity of life, of death standing on our doorstep, ready to take us in. Small stones and dirt on the ground slowly started to rise up into the hole. I could feel the energy, starting to pull me up, as well as gravity reversing itself. I used the remaining force to hold on to a tree while the others, including the General, were pulled up into the hole, their screams were loud, helpless and full of pain. The ground started splitting itself right below me as well as all around me. Everything was coming together, the earthquake splitting up the ground so that the black hole could devour it effortlessly, almost seeming like a well thought out plan of a criminal master mind, such harmony fusing itself with the ultimate apocalypse, all of this manifesting itself while my full sense of awareness was active.

How could this even be possible? I thought, while realizing that it did not matter, since I was being pulled up in the air as well, along with several pieces of terrestrial ground. Looking below me, the sight of the entire world disjoining itself was just to devastating to see, while the sight above me was nothing more than death itself. In the end, I was a failure; in the end I was nothing more than somebody who was unable to fulfill his destiny. No tears were shed by me, how could it end like this, how could I, after everything, not have been destined for something greater than this. In my mind the plea to the gods to have mercy upon my soul was joining with the conviction of the gods having failed to protect the world just as I had, leaving an inexpressible controversial emotion within me. It was I, the Emperor, who was now judging the gods, who was now evaluating their choices.

The energy of the black hole started to push my bones into each other, triggering a physical pain never before experienced, I was being destroyed systematically, while being awake, sensing the pain of my very own death. I could not move a single nerve, and my thoughts were blurry and unclear, I was driving mad, my whole life was spinning in front of my eyes showing images, memories, scenarios, never before seen, melting with the undeniable pain of the black hole crushing my each and every bone, leaving me in a terrorized state of emotional and physical overflow.

Chapter 5

Nothing was there but the white emptiness around me, no floor to stand on, no sealing to look up to. No pain, no joy, no death, no life inside of me, just emptiness, and the awareness of the vacuum space, pure as it was. Where was I? Where was the pain? Where were the Emotions? I could recognize a, with human words, indescribable creature manifesting itself in front of me.

"Who are you, where am I?" I asked desperately.

"You, as entity in this space are nothing and I am Susanoo."

He was it; he was floating in front of me, the cause of the problem, the god of the storms and seas.

"I failed, I failed completely," I said.

"You talk of such nonsense, humans, there all the same, such a young race, relying on such foolish principals," Susanoo stated.

"What do you mean? I am a failure; I failed to fulfill what I was destined to do!" I replied rather upset.

"You speak of failure and destiny, those are nothing more but human notions to separate the truth from reality, do you really think any of that matters now, do you really think this is the process you call destiny, are you really so foolish to still believe that you are a failure for following your common senses of what was right and wrong?"

Upon those words, I was speechless, not being able to understand, not being able to separate myself from my very own reality.

"Life and how you experience it is what humans call reality, it is what they hold on to, it is all that they care about, because of this, they are unable to realize that their so called reality is nothing more than one piece of the puzzle, what you did with your time in that world, was for you to decide, it was never written down for you by anyone, the process you call destiny is a pathetic excuse for humans to relieve themselves from the duty of building their own path. Nothing is written, you did not fail anywhere, nor does it matter now even if you did. While there is always a reason for things being as they are, there is a difference between that and the thing you call destiny."

It was slowly coming to me; in my mind I could feel how my common sense elevated itself from the current realization of reality, to a much brighter and warmer place, a place of peace and safety. I did what I thought was right, I did not take a path against my values and my personality nor did I ever meant to hurt anyone, yet still, many died under my decisions. Here, where I was now, none of that mattered anymore; it was a reality I had to leave behind.

"Were you happy, were you truly happy with your life, with your reality up until now?" Said a different voice from the one of Susanoo.

I could recognize another equally indescribable creature appearing behind Susanoo; it was Amaterasu the god of the sunlight.

"Keeping the good times in focus, yes I was, because I lived with the convincement of myself being a good person, a person with a pure soul," I responded.

"Isn't that what matters now? Looking back at that piece of the puzzle, you were happy, convinced of yourself as a human. Your soul is indeed a pure one, which is why you are here at this time and at this place," Amaterasu said.

"Why did this happen, why did this apocalypse happened?" I asked while I still could.

A third unknown voice, which could then only be Tsukuyomi, the god of the moonlight, answered, "Humans are such an intellectually undeveloped race; there is still so much ahead of them. You could never even begin to understand the motives behind our actions".

Then Susanoo added, "Ultimately, my friend, it was only our reality, our perception of good and evil which led us to act as we did, a perception, a reality, no human could ever understand."

Inside of my soul I was fulfilled, while empty at the same time, I realized that life itself was a small piece of the whole and that death was never the end, but rather a new beginning.

"And what happens now?" I asked the three creatures posing themselves as gods, yet standing in front of me as equals.

"If you still ask such foolish questions, then that only shows that you have not understood much, yet again how could you, human, so young and innocent. We truly wish for you to understand one day, and to live happy, in which ever reality you are going to continue living," Susanoo responded while slowly fading, melting with space and time along with the others right in front of my eyes.

There was not one moment for me to think about the words of the gods, as the undeniable pain returned into my body, the white space started evolving into the pitch black hole again, and the reality I had just came out of started resuming itself in front of my eyes.

It hurt so much, physically and emotionally but ultimately I knew that now, it did not matter, I realized how unbelievably subjective the term reality really was. With that final clear thought in mind I submitted myself freely to the pain, and resumed my process of dying, this time, without fear, without regrets and without destiny, allowing the road ahead to be merle build by non-other than myself.

Epilogue

I felt a light sense of consciousness reclaiming the control over my mind, transporting me from that world to another. My muscles trembled, everything was wet around me, it was raging hot as my consciousness gained full control and I opened my eyes, screaming of the pain in my muscles, lifting my upper body and slowly realizing where I was. I was lying in bed, covered in wet sheets of sweat, feeling as if I just went through hell and back, having tears in my eyes. The pain slowly left, and my muscles relaxed a little. It took me several minutes to fully calm down and realize what had happened. Was it a dream? It had to be, I was in bed and I just woke up. If it was a dream, it was one of the most intense dreams I have ever had in my life.

I could not recall everything as I grabbed some papers and a pencil and started writing down what I did remember. Gods, the apocalypse, death and reality melting and so much more was on my mind. After writing everything down I began to structure the events, realizing the connections between the possible dream and the current world I was in. The gods are Japanese gods we were discussing in college at the time. Pangaea was the supercontinent that existed 250 million years ago, the earthquake was supposed to split it into the continents we know of today, while the Panthalassic Ocean was the name of the super ocean covering the rest of the earth with water. I noticed how some things made sense, while I just could not recall everything that had happened.

It felt so real, everything just felt so real. It was as if I just lived a different life for a night and now I was back here. That very thought, the thought of the undeniable high degree of emotional realism, led me to the conclusion that I could never be sure if I was dreaming or not. I could be dreaming right now, I was in my room, but so was I in my fortress just some hours ago, as somebody completely different. For me, there just was no way to tell.

Ultimately, what makes that dream any less real from what I am experiencing right now? Just because now, I am realizing it as a dream, as something that did not happen in this reality, does not mean that it did not happen in another reality. Who knows, maybe someday, I might realize this reality, the very one that I am living in, as a dream, as something that did not happen in the reality that I will then be in.

At that very moment of intellectual satisfaction, my door opened, I was blinded by the sunrays shooting through my beautiful roof window, soon realizing that it was my sister who laid her arms around my body and pushed it against her own. For some unexplainable reason, I was ecstatic to see her, to hug her; it was almost like one of us had been gone for too long.

About the Author

Daniel Strauss is a writer and artist with an undeniable interest for dreams and their analysis. As of 2014, he attends the Humboldt University in Berlin, majoring in English studies and minoring in Musicology.

He spent his childhood growing up in Central America before moving back to Germany in 2007 to finish school. He speaks fluent English, German, and Spanish. Strauss is an advocate of C. G. Jung's archetypal theories and the collective unconscious. He has had numerous productive as well as beautiful experiences analyzing his own dreams. Archetypal research in relation to archetypal dreams also counts to his interests and continues to lead him to a better understanding of himself and the world.